Intellectual Property Rights and Plant Biotechnology

Proceedings of a Forum Held at the
National Academy of Sciences,
November 5, 1996,
Washington, D.C.

National Research Council

NATIONAL ACADEMY PRESS
Washington, D.C. 1997

NATIONAL ACADEMY PRESS • 2101 Constitution Avenue, NW • Washington, DC 20418

NOTICE: The project that is the subject of this report was approved by the Governing Board of the National Research Council, whose members are drawn from the councils of the National Academy of Sciences, the National Academy of Engineering, and the Institute of Medicine. The members of the committee responsible for the report were chosen for their special competences and with regard for appropriate balance.

This report has been reviewed by a group other than the authors according to procedures approved by a Report Review Committee consisting of members of the National Academy of Sciences, the National Academy of Engineering, and the Institute of Medicine.

The National Academy of Sciences is a private, nonprofit, self-perpetuating society of distinguished scholars engaged in scientific and engineering research, dedicated to the furtherance of science and technology and to their use for the general welfare. Upon the authority of the charter granted to it by the Congress in 1863, the Academy has a mandate that requires it to advise the federal government on scientific and technical matters. Dr. Bruce M. Alberts is president of the National Academy of Sciences.

The National Academy of Engineering was established in 1964, under the charter of the National Academy of Sciences, as a parallel organization of outstanding engineers. It is autonomous in its administration and in the selection of its members, sharing with the National Academy of Sciences the responsibility for advising the federal government. The National Academy of Engineering also sponsors engineering programs aimed at meeting national needs, encourages education and research, and recognizes the superior achievements of engineers. Dr. William A. Wulf is acting president of the National Academy of Engineering.

The Institute of Medicine was established in 1970 by the National Academy of Sciences to secure the services of eminent members of appropriate professions in the examination of policy matters pertaining to the health of the public. The Institute acts under the responsibility given to the National Academy of Sciences by its congressional charter to be an adviser to the federal government and, upon its own initiative, to identify issues of medical care, research, and education. Dr. Kenneth I. Shine is president of the Institute of Medicine.

The National Research Council was organized by the National Academy of Sciences in 1916 to associate the broad community of science and technology with the Academy's purposes of furthering knowledge and advising the federal government. Functioning in accordance with general policies determined by the Academy, the Council has become the principal operating agency of both the National Academy of Sciences and the National Academy of Engineering in providing services to the government, the public, and the scientific and engineering communities. The Council is administered jointly by both Academies and the Institute of Medicine. Dr. Bruce M. Alberts and Dr. William A. Wulf are chairman and vice-chairman, respectively, of the National Research Council.

This report was prepared with funds provided by the Office of Energy Research and the Office of Health and Environmental Research of the U.S. Department of Energy under agreement number DE-FG02-94ER61939.

International Standard Book Number 0-309-05828-7

Additional copies are available from:

National Academy Press, 2101 Constitution Ave., NW, Box 285, Washington, DC 20055
800-624-6242; 202-334-3313 (in the Washington Metropolitan Area)
http://www.nap.edu

Printed in the United States of America

STEERING COMMITTEE

MICHAEL T. CLEGG, *Chair*, University of California, Riverside
ELLIOT M. MEYEROWITZ, California Institute of Technology, Pasadena
RONALD R. SEDEROFF, North Carolina State University

Science Writer

ROBERT POOL, Arlington, Virginia

NRC Staff

JANET E. JOY, Study Director
MARY JANE LETAW, Staff Officer
JULIEMARIE GOUPIL, Project Assistant

BOARD ON AGRICULTURE

*Through December 1996.

COMMISSION ON LIFE SCIENCES

THOMAS D. POLLARD, *Chair*, The Salk Institute for Biological Studies, La Jolla, California
FREDERICK R. ANDERSON, Cadwalder, Wickersham and Taft, Washington, D.C.
JOHN C. BAILAR III, University of Chicago
PAUL BERG, Stanford University
JOHN E. BURRIS, Marine Biological Laboratory, Woods Hole, Massachusetts
SHARON L. DUNWOODY, University of Wisconsin, Madison
URSULA W. GOODENOUGH, Washington University, St. Louis, Missouri
HENRY W. HEIKKINEN, University of Northern Colorado, Greeley
HANS J. KENDE, Michigan State University, Lansing
SUSAN E. LEEMAN, Boston University School of Medicine
THOMAS E. LOVEJOY, Smithsonian Institution, Washington, D.C.
DONALD R. MATTISON, University of Pittsburgh
JOSEPH E. MURRAY, Wellesley Hills, Massachusetts
EDWARD E. PENHOET, Chiron Corporation, Emeryville, California
EMIL A. PFITZER, Research Institute for Fragrance Materials, Inc., Hackensack, New Jersey
MALCOLM C. PIKE, Norris/University of Southern California, Comprehensive Cancer Center, Los Angeles
HENRY C. PITOT III, University of Wisconsin, Madison
JOHNATHAN M. SAMET, John Hopkins University, Baltimore
CHARLES F. STEVENS, The Salk Institute for Biological Studies, La Jolla, California
JOHN L. VANDERBERG, Southwest Foundation for Biomedical Research, San Antonio, Texas

NRC Staff

PAUL GILMAN, Executive Director
SOLVEIG M. PADILLA, Administrative Assistant

Preface

In 1993 the National Research Council's Board on Biology established a series of fora on biotechnology. The purpose of the discussions is to foster open communication among scientists, administrators, policymakers, and others engaged in biotechnology research, development, and commercialization. The neutral setting offered by the National Research Council is intended to promote mutual understanding among government, industry, and academe and to help develop imaginative approaches to problem solving.

For the first forum, held on November 5, 1996, the Board on Biology collaborated with the Board on Agriculture to focus on intellectual property rights issues surrounding plant biotechnology. It was suggested that plant biotechnologies have not developed with the same vigor as might have been expected, given recent progress in molecular biology and by comparison to biomedical biotechnology. It was hoped that a forum could clarify intellectual property issues among research collaborators and potential impacts on advances in plant molecular biology.

Participation at the "Forum on Intellectual Property Rights and Plant Biotechnology" by representatives of the U.S. Department of Energy, U.S. Department of Agriculture, National Science Foundation, National Institutes of Health, and U.S. Patent and Trademark Office (PTO) suggests that intellectual property rights issues are important to many federal agencies. Forum participants agreed that exploration of successful technology transfer models would benefit public-private research collaborations. Executives from biotechnology firms raised concerns that universities overestimate the value of intellectual property in the marketplace. Scientists emphasized the need for a research exemption to prevent intellectual property rights from negatively impacting the research environment.

Examination of intellectual property rights also has been a focus of other activities of the Board on Biology. In 1993 Harold Varmus, chair of the Board on Biology at that time, headed an effort to discuss sharing of reagents associated with transgenic mice. More recently, the board organized a forum in November 1995 to examine the effects of intellectual property protection on the development, dissemination, and utilization of research tools such as expressed sequence tags and the polymerase chain reaction. It is anticipated that the present forum proceedings will generate further interest in intellectual property rights and other issues of biotechnology.

Michael T. Clegg, *Chair*
Board on Biology

Contents

1

Overview

Molecular biology has both transformed the science of biology and spawned the new industry of biotechnology. Plant biotechnology is a knowledge-based industry that depends on basic research tools and novel genetic combinations for continuous innovation. In efforts to increase U.S. competitiveness in the 1980s, federal policies were implemented to encourage public-private research collaboration and to promote more rapid commercialization of new inventions. Today, in exchange for ownership of enhanced germplasm and scientific knowledge, industry is supplying its university collaborators with much-needed research funding. In addition, the plant biotechnology industry has gained strong protections for its innovations through the granting of utility patents. Academic scientists are uncertain what effect this strengthening of property rights will have on plant molecular biology in the future. Sorting out claims to ownership of intellectual property is perhaps the most difficult issue facing universities and industry as they strive to create new partnerships.

In November 1996 the National Research Council convened a forum of scientists and administrators from universities, industry, and the federal government to explore intellectual property rights issues from the diverse areas of plant science that support development of future plant biotechnologies. The purpose of the "Forum on Intellectual Property Rights and Plant Biotechnology" was to promote an open exchange of views among research collaborators in crop genetics, phytoremediation, and biobased energy sectors. The present volume summarizes the discussions and issues raised by participants at the forum.

Increased interest in the protection of intellectual capital has stimulated the establishment of technology transfer offices in universities across the United

States. Though forum participants indicated that more and more scientists are aware of the need to protect their inventions, university researchers need better guidelines for balancing patent and publishing rights. Some participants noted that universities have an unrealistic expectation that immature technologies will be developed rapidly into useful products by commercial firms. Forum participants agreed that the generation of guidelines for successful technology transfer would improve relationships among research collaborators.

Forum participants represented a continuum of plant science applications ranging from the mature hybrid seed industry to small start-up biotechnology firms. Since 1990 most small genetic engineering firms that developed food and fiber crops have been acquired by larger firms. On the other hand, small biotechnology spin-offs in the phytoremediation sector are beginning to attract the attention of venture capitalists. Participants explained that intellectual property is a strategic tool to increase a firm's competitive advantage. According to those engaged in biobased energy research, since private corporations anticipate eventual returns on their investments, federal funding will continue to be a major component of biobased energy research until marketable products are developed. Forum participants agreed that all three segments of plant science will increasingly depend on industrial funding for basic research underlying the development of biotechnology innovations.

At the forum, academic scientists expressed uncertainty about the effect that strengthened intellectual property rights may have on the future of fundamental plant science. Some argued that strategies developed to protect ideas and data are inhibiting an open laboratory, which is so vital to the discovery process. Others emphasized that access to enabling technologies and genetic material now concentrated in some private firms is crucial to the improvement of most food and fiber crops. Ron Sederoff, speaking for the majority of forum participants, warned that intellectual capital in plant molecular biology is deteriorating. Without increased public funding resources, universities cannot create new knowledge or train scientists. If this trend continues, universities will have little intellectual capital to offer industry. As a result, innovations in plant biotechnology will suffer. Forum participants agreed that government, university, and industry collaborations will benefit from continued exploration of intellectual property rights issues.

2

Summary of Forum Proceedings

WORLDS IN COLLISION

Over the past 25 years plant research has undergone a revolution. In 1970 it was a discipline that, while important, moved relatively slowly and seldom made headlines. Today, powerful techniques developed in plant biology have the potential to transform plants into valuable commodities for the agricultural biotechnology community. With this success has come a growing tension between academia and industry over how knowledge in this quickly changing field should be generated and shared. Academic scientists worry that commercial interests will skew the direction of their research and crimp their traditionally open communications, while businesses contend that scientists fail to appreciate the importance of intellectual property rights.

In modern agricultural research this conflict has taken on an unusual urgency. The reasons for the urgency can be found in three parallel developments that have profoundly shaped plant research and development. The first is the ongoing revolution in molecular biology and the incredible power that it has placed in the hands of plant researchers. Equipped with the tools of genetic engineering, scientists can introduce new traits into plants far more quickly and surely than can be done with traditional breeding methods. More importantly, the traits can be things, such as resistance to certain herbicides, that are all but impossible to create with traditional techniques. The result is a new field—plant biotechnology—that is completely transforming the way plant science is performed.

The second factor is that new federal patent legislation, court decisions, and

BOX 1 Plant Protection

Over the past quarter of a century, Congress and the courts have greatly strengthened the protection of intellectual property rights for biological inventions, including plants. Below is a brief history of the rights for plant inventions:

The original Patent Act of 1790 provided no protection for plants or animals, no matter how much intellectual effort had gone into producing a particular variety or breed. Plants and animals were considered to be "products of nature" and thus not patentable.

In 1930 Congress passed the Plant Patent Act, which allowed the granting of "plant patents" for asexually reproduced plants—those that are reproduced by means other than seeds, such as by tissue culture or propagation of cuttings. Asexually reproduced plants, which are genetically identical to their donor plants, include many types of fruit and nut trees and also ornamental plants. The act did not include protection for sexually reproduced plants because at the time it was thought that plants grown from seed could not be guaranteed to be identical to their parents. The act also excluded tuber crops.

Forty years later Congress provided a different sort of protection to sexually reproduced plants other than hybrids with the Plant Variety Protection Act of 1970. By this time it was clear that plants grown from seed could remain true to type from generation to generation, so the act allowed the U.S. Department of Agriculture to safeguard new varieties that were stable, distinct, and uniform by issuing Plant Variety Protection Certificates. The protection offered by these certificates, however, was relatively weak. Only exact copies were covered, so a breeder could introduce minor cosmetic changes in a variety and get a separate certificate. Furthermore, the owner of a protected variety could not prevent other breeders from

rulings by the U.S. Patent and Trademark Office (PTO) have greatly strengthened and broadened the intellectual property protections available for biological inventions. In the 1980s, for example, it first became possible to obtain a utility patent—the strongest patent in terms of the level of protection it offers—on new types of plants or animals, whether created by traditional breeding practices or genetic engineering. The increased protection for intellectual property may have made industry less likely to keep some inventions as trade secrets, but it has also given inventors greater power to control access to inventions that are made public.

The third factor is a new attitude toward patenting federally funded research, produced by passage of the Bayh-Dole and Stevenson-Wydler Acts in 1980 and the 1986 Technology Transfer Act. Up through the 1970s, although there were exceptions, research performed with money from the federal government was usually put into the public domain. Now, however, universities, government agencies, and the researchers working at those places are encouraged, even expected, to file for patents on commercially valuable inventions as part of a push to using the plant in their own breeding programs. Nor could the owner of a variety keep farmers from saving seeds for their own use or to sell to others.

This protection was upgraded in the 1994 Plant Variety Protection Act Amendments. Now the protection certificates guard against "essentially derived varieties," which are varieties that differ from the protected plant by only minor changes, although Congress was rather vague on what differentiates "minor" from "not minor" changes. Farmers must now get a license to sell seeds of protected varieties, although they may still keep the seeds for their own replanting. The 1994 amendments also extended protection to tuber crops and first-generation hybrids.

Despite this string of laws expanding protection, the most significant change in intellectual property rights for biological inventions did not come from Congress. It came instead from the U.S. Supreme Court and the U.S. Patent and Trademark Office. In the landmark 1980 case *Diamond v. Chakrabarty*, the court ruled that a genetically engineered microorganism could be patented under the 1790 Patent Act. Such an organism meets the criterion of a "manufacture" or "composition of matter," the court held. Following this ruling, the U.S. Patent and Trademark Office extended the reasoning to plants and animals in a series of rulings during the 1980s. Now utility patents—the type of patent created in the 1790 act—can be awarded for new types of plants, including seeds, plant parts, tissue cultures, and plant genes and also for new breeds of nonhuman animals.

Although it is more difficult and expensive to receive a utility patent than a plant patent or a Plant Variety Protection Certificate, the utility patent provides much stronger and broader protection. The standard is obviousness, and the test for obviousness is whether the claimed subject matter would have been obvious to a person of ordinary skill in the art at the time the invention was made. This is a legal determination. Reasonable expectation of success goes to motivation or rationale as to why one of ordinary skill in the art would have found the claimed invention obvious. If the invention as claimed is not obvious, an innovation can be patented.

get a greater return on publicly funded research. The theory is that businesses will be more likely to develop marketable products from a university or government agency's inventions if the inventions are protected by patents. One result of introducing intellectual property considerations into the academic setting, though, has been to upset long-standing traditions about how the results of scientific research should be disseminated.

The effect of these developments has been to bring two quite different worlds into collision: the academic world, with its emphasis on open inquiry and free exchange of ideas and information, and the business world, with its eye on the bottom line and its resulting insistence on guarding—and sometimes hiding—its most important information. Because plant biotechnology is so new and is evolving so quickly, businesses find themselves much more reliant on basic research being performed in university and government laboratories than is usual. Conversely, researchers in university and government laboratories find they must ask private companies for permission to use basic tools they need to do their jobs. Because the field is so young, the patents on its major tools—genes, promoters,

BOX 2 Patenting Federally Funded Research

Twenty years ago universities and government agencies seldom patented the results of research they performed with federal funding. There was a sense, particularly in land-grant universities, that research made with public funding should be placed in the public domain. But a series of three federal acts changed that attitude dramatically: the Bayh-Dole Patent Policy Act (1980), the Stevenson-Wydler Technology Innovation Act (1980), and the Federal Technology Transfer Act of 1986. The Bayh-Dole Act allowed individual researchers to patent and grant licenses for research done with federal funds; it also encouraged universities to make sure their intellectual property was properly developed. The Stevenson-Wydler Act and its subsequent amendment, the 1986 act, set up CRADAs (cooperative research and development agreements) as a mechanism for collaborations between government and private research laboratories and directed federal agencies to transfer technology to private firms. These three acts have pressed academia and government to join the intellectual property rights game.

That development has not left everyone happy. Several forum participants argued that inventions made with public monies might best be left to the public. One was Suzanne Scotchmer, an economist at the University of California, Berkeley. "Since this is government-funded research," she asked, "why isn't it better just to put it in the public domain without applying for intellectual property rights on it?"

Responding to such concerns, June Blalock, a licensing specialist with the Office of Technology Transfer, U.S. Department of Agriculture, reminded the participants that the Bayh-Dole Act was passed primarily because the public was not benefiting from public domain inventions made with federal funding due to the absence of an incentive for the private sector to commercialize unprotected inventions.

One reason for that, noted George Jen, an associate at Pennie and Edmonds LLP, is that typically a great deal of development is needed to turn an invention into a valuable product, and businesses are reluctant to pay for such development unless they know their investment is protected by intellectual property rights.

This was the experience, for instance, at Iowa State University, said Patricia Swan, vice provost for research and advanced studies. In 1990 the university began patenting its inventions in the seed crop area, but "had the experience just prior to 1990 of realizing that most of the crop varieties that we were offering under public release—the traditional way of releasing crop varieties for almost a hundred years—were simply not being picked up by farmers." Two of the university's largest breeding programs were for corn and soybeans, but nearly 90 percent of the soybeans and more than 95 percent of the corn grown by Iowa farmers was being provided by commercial breeding programs. The private companies attained their market share, she said, by marketing their seeds well, putting together excellent distribution systems, and providing farmers with services and other needed products along with seeds—but they put these development efforts into seeds for which they hold the intellectual property rights, not for seeds in the public domain.

In at least some cases, however, technologies in the public domain do get used, said Alan Bennett, associate dean of plant sciences at the University of California, Davis. "Unlike Patricia Swan's example, where the public germplasm was not being adopted, we have found in California that much of the public germplasm is adopted because there really are [no] commercial alternatives [for] many of these crops." California agriculture, he noted, is dominated by minor crops, which attract far less commercial breeding interest than corn or soybeans.

enhancer sequences, transformation systems, markers, and so on—are still in force. Furthermore, even those tools developed with federal grants by researchers at other universities are often not freely available but must be licensed at a certain cost and inconvenience.

Meanwhile, the ability to patent federally funded work has changed the relationship between private businesses and academic and government institutions doing commercially valuable research. The public institutions must now worry about such things as what to charge for licenses to their inventions and how to divvy up rights to research done jointly or with funding from industry, and the researchers at those institutions have become more like researchers in industry when it comes to protecting their intellectual property rights. More importantly, possible manifestations of this trend might ironically have a negative long-term impact on growth of the research and innovation base upon which future industry will in fact depend.

The ultimate success of plant biotechnology will depend in large part on how well universities, businesses, and government agencies can learn to work together through collaborations and consortia, licensing arrangements, technology transfer contracts, cooperative research and development agreements (CRADAs), and the like. To assess how well these goals are being met, the National Research Council assembled several dozen representatives from academia, industry, and government to participate in its "Forum: Intellectual Property Rights and Plant Biotechnology" in November 1996. The forum participants were asked to address such questions as:

- Do current means for protecting intellectual property rights adequately encourage both scientific progress and commercial development?
- Is technology transfer managed in a way that provides for scientific progress, incentives for commercial development, and public benefit?
- What benefits and problems result from negotiations or alliances between university, government, and commercial laboratories?

The following is a distillation and synthesis of the presentations and discussions at the forum.

WORRIES IN THE LABORATORY

Although researchers in agricultural biotechnology are excited to be working in such a dynamic field and happy that their efforts often have valuable applications, they are not so pleased with what they see happening to the academic environment. The emphasis on the commercial value of research and on protecting intellectual property rights is affecting how basic research is done at university and government laboratories, and many scientists worry that the changes are for the worse.

One major concern centers on the availability of research tools. Access to some tools, such as the basic technique of transferring DNA developed by Stanley Cohen and Herbert Boyer in the early 1970s, is straightforward. "That patent is quite clear and available to all," noted Alan Bennett, associate dean of plant sciences at the University of California-Davis, at the forum. Stanford University, which holds the Cohen-Boyer patent, offers relatively inexpensive licenses to any institution wishing to do genetic engineering research. But, Bennett said, the situation is trickier for many of the other tools that biotechnology researchers need: "In the case of such enabling technologies as *Agrobacterium*-based gene transfer, certain promoters, selectable markers, and gene suppression technology, it is less clear—and, in many cases, quite uncertain—how access to those enabling technologies can be achieved."

Consider, for instance, the case of *Agrobacterium tumefaciens.* This microorganism, a pathogen that causes tumors in certain plants, has the natural ability to insert its own genes into plant cells. Scientists have adapted this ability for their own purposes, using *Agrobacterium* to ferry genes from various organisms into plants. Today, although other techniques are available, *Agrobacterium*-based gene transfer is considered the most widely useful method of modifying a plant's genetic makeup. Unfortunately, just who will own the rights to use *Agrobacterium*-based gene transfer in the United States is up in the air right now. Several institutions have filed conflicting patent applications, and the U.S. Patent and Trademark Office is now working to settle the matter. Until then no one can be sure which party to negotiate a license with, so the commercial rights to any plant modified with *Agrobacterium* are uncertain.

Even when the ownership of a technology is not in doubt, academic researchers sometimes find they are shut out from using inventions whose rights are controlled by private companies. At Iowa State University, for example, plant breeders have been rebuffed a couple of times when they approached a company about licensing a technology. "We were refused, even though the company is licensing to many other companies," said Patricia Swan, vice provost for research and advanced studies at Iowa State University. "The company indicated that [it] did not want to license to us because [it] did not believe that universities were capable of managing and looking after the intellectual property in the way that it should be looked after."

Researchers at government agencies face the same problem, said Robert Swank, director of research at the U.S. Environmental Protection Agency's (EPA) National Exposure Research Laboratory in Athens, Georgia: "Not all companies and not all universities are very free in giving us their proprietary information, even in a research domain. In effect, we operate in a research-exemption mode. Everything we do is yours. But the converse of that is not true, and it does hamper our ability to conduct research."

This is especially true in those areas of agricultural biotechnology that are so new it is difficult to judge the eventual size of the market. At Oak Ridge National

Laboratory, Gerald Tuskan, a forest geneticist, is working to develop fast-growing trees that can be biochemically converted to liquid transportation fuels. The U.S. Department of Energy, as part of a national program to develop renewable energy sources, would like to see 200 million tons of such biomass produced each year by the year 2020, which will require between 20 million and 40 million acres to be planted with trees or other energy crops. This would be one of the country's largest single-commodity crops, but today the size of the crop is zero, and that creates a problem, Tuskan said. "It's hard to negotiate with Monsanto or Ciba-Geigy or other biotech companies to get the rights to gene constructs for a crop that has no value base in today's terms but has the potential to be rather large."

On the other hand, Robert Fincher, director of university collaborations and germplasm licensing for seed industry giant Pioneer Hi-Bred International, offered a more sanguine assessment of the availability of technology to researchers. Despite the complaints from universities and government, he said, in his experience industry refusals to share technology are the exception rather than the rule:

> Germplasm [the genetic material of plants] and biotechnology inventions can usually be obtained from companies through research or license agreements. A few times, friends of mine in the plant-breeding community at universities have noted in a newsletter or industry journal that Pioneer received a patent on such and such, and they have drawn the conclusion, "Well, that means the door is shut to us." That is not true at all. They need to avoid jumping to those conclusions and contact our technology transfer office and find out how they can access the invention because usually it is available for research purposes.

Fincher did acknowledge that there will be times when companies, for one reason or another, decide not to make a patented technology available. In such cases, researchers must wait for the company to change its mind, find another way to attack the problem, or move on to something else.

Sometimes the shutting out of researchers from a technology or line of inquiry is less direct but no less effective. Bennett described one such conundrum in California. As part of a project funded by the Strawberry Commission, researchers had been working to insert a gene into strawberries that would cause the berries to produce fungus-killing chemicals and so reduce the need for fungicides. Researchers were using an antifungal gene and a strawberry cultivar both patented by the University of California, so access to them was no problem. Unfortunately, however, as the project progressed, those involved realized that access to other necessary technologies—*Agrobacterium*, to insert the gene, promoters, and selectable markers—was not nearly so clear. Indeed, Bennett said, it appeared that even if the researcher succeeded in developing a strawberry line with antifungal properties, difficulties in getting commercial rights to the various technologies would make it impossible to market the line. The Strawberry Commission dropped its funding of the program.

"We now find that this is rippling throughout many commodity boards,"

BOX 3 A Research Exemption?

Several members of the forum panel called for a research exemption—a policy making patented inventions available for use in pure research—as a way of guaranteeing that scientists have access to the tools and materials they need. Without such an exemption, Ronald Sederoff, geneticist at North Carolina State University, said researchers face the fate of Nobel Laureate Arthur Kornberg, who "devoted his life to working on DNA polymerase yet is precluded from purifying Taq polymerase himself and using it for PCR [polymerase chain reaction] because of the structure of the patents that have been used to protect PCR."

Kornberg discovered DNA polymerase in the 1950s, and it was only by building on his work and that of many others—all of whom put their research in the public domain—that Kary Mullis of Cetus Corporation was able to create the now-famous PCR. Yet Cetus patented that 1985 discovery and the heat-stable *Taq* polymerase that made it possible, and now anyone who wishes to use *Taq* polymerase in PCR must get a license to do it—by buying their *Taq* polymerase from Hoffmann-La Roche, which bought Cetus's PCR patent rights in 1990. "It seems inappropriate," Sederoff said, "that all the people who invested in researching polymerase before Mullis [are] not allowed to use PCR even for noncommercial purposes." Setting out a workable research exemption is much harder than it sounds though. In theory, for instance, a research exemption already exists, but it is extremely narrow. Traditionally, judges have accepted that scientists and others cannot be barred from using patented material purely for "philosophical purposes." If, however, there is some commercial value to the research, it is subject to control by the patent holder, and the line between the purely philosophical and the commercial is not at all clear—assuming that such a thing as "purely philosophical" even exists. In practice, judges have almost never allowed a research exemption when the issue came before a court, but the cases that do come before a court are generally those in which the patent holder feels threatened, and those will be ones in which some commercial interest is at stake. If a researcher were, for example, to experiment with a patented material purely to explore how it worked, the patent holder would probably not file a suit to stop it, and, if such a suit were filed, a court might well rule for the researcher.

Still, observers have suggested several ways that the current patent system could be changed to assure scientists of freer access to the things they need in their research. One way would be for Congress to change the patent law to include exemptions carefully delineated to protect commercial prospects while allowing basic research to move forward. Recently, Congress did just that for the pharmaceutical industry, giving researchers the right to perform trials of a patented drug before the patent lapses in preparation for later sale of a generic form of the drug. Before passage of that legislation, a court had ruled that such research did infringe on the drug's patent.

Another possibility, according to John Barton, a professor of law at Stanford University, would be legislation requiring that inventors of research tools developed with federal funding offer royalty-free licenses to other researchers who needed those tools. So far, however, discussions of changes in the research exemption have not gotten past the talking stage, and no one can predict what, if any, help researchers may get.

Bennett said. "It is affecting their willingness to support research in the genetic engineering of minor crops because of the uncertainty as to how things can reach the commercial market. Until we find some path to access enabling technologies, participation in public research programs on this direct application of genetic engineering is effectively on hold."

Just as worrisome to researchers as limits on access to technologies is what they see happening to the traditions of openness and unfettered communication in science. Most scientists believe that research in every area of science has been most effective and has progressed most quickly when scientists have been able to talk freely with their peers, exchanging ideas and hypotheses, passing along new techniques, describing their latest results, and debating what it all means. But the need to protect intellectual property rights is putting a damper on all this.

The best-known example is how patenting concerns slow the publication of scientific results. Steven Strauss, a forest geneticist at Oregon State University who leads a university-government-industry consortium seeking to develop transgenic trees for commercial use, such as pulp and paper or biofuels, described the situation this way: "Most of the licensing agreements we get, even on a research-only basis, try to censor, delay, or bias publications that will come out of privately funded research that we do. So we have fights that often take months or years to resolve." That slows progress in the field and also has serious practical implications for the researchers themselves. "The consequences," Strauss said, "are most serious for students, postdocs, and young professors, for whom publication in a timely manner is critical to career development. If you publish something quickly, you may get into a prestigious journal. If you publish it slowly after other key publications might have come out, you may get into more of a trade journal, with large consequences for the prestige and recognition that you get."

The situation is similar for researchers in government agencies, indicated Robert Swank. "We are starting to see delays in publication—one to three years is not uncommon—while we work out arrangements with all the parties, public and private, who may be involved."

But the pressures on scientific communication are much more profound and more troubling than simply delays in publishing in peer-reviewed journals, said Ronald Sederoff, a geneticist at North Carolina State University. To patent an invention or discovery, one cannot disclose the work before applying for the patent, and "disclosure" is a very broad term. "Every time anybody steps out of a laboratory and says anything or writes anything that is not covered by a confidentiality agreement, it is a potential disclosure," Sederoff explained. Until a patent application has been filed, researchers not only must refrain from publishing reports of their work but should also avoid discussing their work at professional meetings, departmental seminars, and perhaps even informal meetings with colleagues down the hall. According to Sederoff, "It means the end of something I have treasured all my life: the open laboratory."

Ironically, some of the pressure to restrict communication comes not from the companies funding the research, Sederoff said, but from university and government agency administrators seeking to protect their institutions' intellectual property rights. "A significant burden has been placed on us by the government and through the Bayh-Dole Act." Universities and government agencies have traditionally been committed to open scientific communication, but now that they have a stake in commercializing their researchers' inventions, they are imposing on their researchers many of the same censures that industry has long imposed on its employees.

The financial stakes and the ability to patent are also, Robert Swank said, having an effect on individual researchers, whom Bayh-Dole allowed to get royalties from the licensing of patents. "Until recently we were public servants first. Our idea was pure discovery and publish. Put it out there, let the regulators and the industries use it however they want, and we would stand back and just provide consultation services. That is no longer the way it works."

Instead, Swank said he is beginning to see scientists looking out for their own financial interests. "Researcher A says, 'I can make a million bucks off of this bright idea . . . I just had. Am I going to work with my fellow researcher down the hall and share that? No way.' " Because of that attitude, Swank said, teamwork at the U.S. Environmental Protection Agency (EPA) is suffering.

Finally, many scientists also worry that the pressures for research to have short-term payoffs will push science away from the emphasis on long-term basic research that made the genetic engineering revolution possible in the first place. This is particularly troubling because federal funding for basic research has not been increasing as much as in the past. Patricia Swan summed up the worries this way: "Some of our innovations come out of 20 years of poking around inside a plant and trying to figure out how something is working. A biotech start-up company is not going to be able to take that kind of risk. So the question today is if our breeding programs are going more toward the immediately useful rather than the poking around inside of the plant and seeing what is going on. "

COMMERCIAL CONCERNS

While the patent system may seem quite restraining to scientists in university and government laboratories because it forces them to put off communicating their work longer than they otherwise would, it actually serves as an agent for openness for researchers in the business world. If a company wishes to have an invention legally protected by the patent system, it must make the details of that invention part of the public record.

In short, protecting their intellectual property is an overriding concern for companies in the agricultural biotechnology field, and this shapes their perception of how intellectual property rights are affecting their industry. While forum representatives from universities and government agencies worried about how the push for intellectual property protection might damage the research environment, participants from industry emphasized how important that protection is for them. It was also clear, however, that intellectual property rights are important to different companies for different reasons. In particular, large firms and small firms use intellectual property in very different ways.

Large agricultural companies make much of their money by selling products to farmers. For them patents and other forms of intellectual property protection are most important in guaranteeing that they will secure a large market share on the end products for which they spent a great deal of time and money to develop. "Patenting germplasm and biotechnology inventions is critical to our ability to deliver useful products and get paid for those products," said Robert Fincher of Pioneer Hi-Bred International. "We do not want to have to compete against our own inventions and investments, which has happened in the case of germplasm prior to the development of good intellectual property protection for varieties and hybrids." By using patents to prevent competitors from selling identical products, Pioneer and other companies can make a good return on their investment in research and development.

Small companies see intellectual property in a different light. As was pointed out at the forum, smaller biotechnology companies, particularly start-ups, may have few or no products that they sell to customers. Instead, their value lies mostly in their intellectual property, and they use patents to raise money. With patents in hand, a biotechnology firm can go to investment bankers or capital markets and get funding for research and development of products that are still far in the future. One of the most dramatic examples of the market value of intellectual property was last fall's purchase of PGS International by Hoechst Schering AgrEvo. PGS's assets were worth only $30 million or so, but AgrEvo paid $730 million for the company. This meant AgrEvo valued PGS's intellectual property—which included technology for making plants resistant to a broad-spectrum herbicide made by AgrEvo and for using the *Bacillus thuringiensis* (Bt) toxin gene to make plants resistant to insects—at more than 20 times its "real" assets.

Because intellectual property is of such overriding importance to small companies, a firm's agenda and strategies are often dictated by patenting considerations, John Bedbrook, executive vice-president of DNA Plant Technology Corporation (recently acquired by Empressas La Moderna) told the forum. "First, patent strategy is important in terms of how it impacts the development of science within organizations. If you want to patent something, you have to go about doing the experiment in a way that leads to a patentable invention, so it has a significant impact on your research design. It also brings a certain discipline to an organization that as a university researcher you do not typically have."

Furthermore, Bedbrook said, because being first to file a patent is so important, "speed and secrecy are as important as the quality of research in this business. I think that has a significant impact on the overall intellectual merit of what

BOX 4 AN ECONOMIST'S VIEW

Suzanne Scotchmer
Professor of Economics and Public Policy, University of California, Berkeley

In the 1980s the Patent and Trademark Office began granting broad utility patents on plant-related inventions. Some patents cover bioengineering tools, such as vectors for transferring genetic material, and in several notorious cases the broad patents have covered bioengineered crop seeds, with very broad claims extending the patent to other types of bioengineered resistance in the same crop seeds and to other crop seeds with similar resistance. This has led to controversy over the proper extent of protection.

Economists view patent protection as an incentive instrument: the objective is to ensure that firms can recover their costs of invention. However, there are conflicting views of whether this objective is helped or hindered by the recent broad patents on crop seeds. Such patents look very good to the first patent recipient, who sees profit opportunities in either developing related products or granting licenses to other inventors. They are viewed less favorably by firms that are shut out of the research enterprise by the threat of patent infringement; for example, those genetic engineers who might want to use the first firm's technology as a guide for modifying their own crop seeds or modifying the first patent holder's crop seed to resist a new pest.

These two views and their merits can easily be illustrated by an example. Suppose a firm discovers how to make a first-generation pest-resistant seed, which has market value F (for first), and suppose that the R&D cost for the pest-resistant seed is $c(F)$. The firm hires a patent attorney who obtains claims as broad as possible. Suppose that a second-generation firm then uses this technology to introduce pest resistance into another seed, with market value N (for next), and incremental R&D cost $c(N)$.

The second seed either infringes the initial patent or not, depending on how broad the first patent is. (This is independent of whether the second-generation seed is itself patentable—it can be both patentable and infringing.) The question is, how does infringement (licensing) affect the profitability of the two generations, and which of the two generations gets invented? Which of the above two views is correct?

is going on in a small company." Finally, he noted that companies will often follow a particular line of research even after it's clear that the research doesn't fit into their overall strategy simply because of the value of the intellectual property. "Frequently, companies will pursue a research objective that was initiated in good faith under the auspices of the overall business strategy, but ultimately pursue it because of its intellectual property value as trading chips," said Bedbrook. The company will not use the research itself but will use it to barter with other companies for rights to their inventions.

Of course, large companies do many of the same things in response to intel-

CASE 1

Suppose that the first patent is broad so the second seed infringes. After it is developed, the second inventor must license the first patent. Suppose the license fee is L. Then the profit of the first innovator is $F + L - c(F)$, and the profit of the second innovator is $N - L - c(N)$. It might occur that $N - L - c(N) < 0$ even if $N - c(N) > 0$. The latter means that the second product "should" be invented. However, if the second firm anticipates that it will lose money because of the high licensing fee, it will not invest. This is the danger of broad patents.

One might ask why the licensing fee L would be so high as to render the second seed unprofitable. The fee is negotiated after the R&D costs $c(F)$ and $c(N)$ are both sunk, so the R&D costs should not affect the licensing negotiation. To some degree this problem might be remedied if the two firms can strike a deal before sinking the costs.

CASE 2

Suppose that the first patent is narrow so the second seed does not infringe. Then the first patent holder earns no royalties from the second seed; their profits are respectively $F - c(F)$ and $N - c(N)$. However the first seed is typically more costly to invent ($c(F) > c(N)$). It might be the case that $F + N - (c(F) + c(N)) > 0$ but $F - c(F) < 0$. In that case the first seed would not be developed, which would also block the second seed from being developed. This is so, even though the two seeds together would make positive profit. That is the danger of narrow patents.

Thus broad patents might stymie the second generation, but narrow patents might stymie the first generation as well. Which type of protection is better depends on the relative costs. If the first seed is very costly relative to its market value, that is, $c(F)$ is larger than F, then a broad patent might be warranted, as a means to transfer profit from the next generation of seeds. But if that is not the case, there is no reason for a broad patent. It is not required to elicit investment in the first seed product and, worse, might stifle the second.

Since there is no provision in patent law to let the breadth of claims depend on the cost considerations outlined here, this reasoning provides a basis for funding basic research.

lectual property considerations, but the effect may not be so great. Most large companies do not live and die by the value of the intellectual property.

With regard to access to enabling technologies, the business world has much in common with universities and government agencies. Everyone recognizes that without access to the basic tools and materials of agricultural biotechnology no progress is possible.

This concern about access to enabling technologies is quite new to the world of agricultural research, Robert Fincher of Pioneer Hi-Bred International noted. "Prior to the development of biotechnology, there was little to consider about intellectual property in the seed industry. You just developed the variety of the hybrid and put it on the market. Now with all of the many components of technology that enter into an improved crop—a gene, its promoter, enhancer sequences, the transformation system, and the selectable marker, all of whose rights may be held by different institutions or different companies—we have to go through a product clearance process where we look at what is in the product and if we have the right to put it on the market. Are there components that we need that we do not have ownership of? If so, we seek a license. If a license cannot be obtained, we look for other technologies and solutions."

In theory, small firms do something similar, but in practice it's much more difficult. John Bedbrook said, "Pioneer can basically sit down and plan a strategy for obtaining all the components in any one activity [it is] involved with. But it is a lot easier to do that when you are sitting on a billion-dollar seed business than when you are a struggling small company trying to convince investors that you are moving toward commercialization of the technology they have invested in."

Indeed, Bedbrook said, "freedom to operate" is the key factor that small biotechnology companies worry about. They must have access to the tools and materials they need to pursue their research, and they have much less flexibility and ability to go around obstacles than do large companies. Although small firms in many areas of business have similar worries, the situation in agricultural biotechnology today is "probably unique in the history of technology-driven businesses," he said. "As the technology is becoming a commercial reality, we have no clear access to the intellectual property that is fundamental to much of our research."

The threat to access comes in two forms. The first consists of monopolies on certain tools and techniques granted by the patent office and the courts. One of the best-known examples is the Agracetus patent on transgenic cotton, which gives Agracetus control over all genetically engineered cotton, no matter what technique is used to transform it. Such monopolies shut out companies from certain lines of research. Agracetus, for example, will not allow other companies to do research on modifying the cotton fibers strengthening them, since it wants to keep that potentially lucrative application for itself; the company does, however, make licenses available for other research purposes.

Perhaps even more frustrating than monopolies are situations in which patent rights are not clear because they are being fought out in court. "The biggest problem for businesses right now is what I call the liability of uncertainty," Bedbrook said. "We live and we operate in an age where certain patents are in dispute and we have to take significant risks because the outcomes are not going to be known for some time." Mentioning the *Agrobacterium*-based gene transfer patent dispute—"the mother of all interferences"—Bedbrook guessed that it could take another four or five years to resolve. And, since patents in the United States extend for 20 years from the date of issuance, the ultimate winner of that patent dispute could control access to a good chunk of agricultural biotechnology until 2020.

The flip side of the coin, Bedbrook said, is that any company that has devel-

BOX 5 Patent Stumbles

Forum participants offered a number of concerns about how patents are awarded in the area of agricultural biotechnology:

Some patents are too broad. In 1992, for instance, a small biotechnology company, Agracetus, was given the patent rights to all genetically engineered cotton made in the United States by any means whatsoever, although Agracetus had inserted a foreign gene into cotton by one particular method, using the gene gun (biolistics). Now Agracetus, owned by Monsanto, has control over all transgenic cotton in this country, a situation that, some worry, could slow cotton research.

Patents are granted for innovations that seem obvious to people in the field. Steven Strauss, a forest geneticist at Oregon State University, called these "me too" patents. Many patents have been awarded, he said, for genetic manipulations of trees that are nothing more than the application of generic techniques borrowed from crop plants. "Trees are really just big plants. Unless there's something very novel in the techniques and the tools, those kinds of patents just add costs and complexity. They're not in the spirit of what patents are supposed to do."

In response, Mary Lee, deputy director of the Biotechnology Group, U.S. Patent and Trademark Office, said that whether something is "obvious to try" is not the standard by which patents are judged. "The standard is obviousness, and the test for obviousness is whether the claimed subject matter would have been obvious to a person of ordinary skill in the art at the time the invention was made. This is a legal determination. Reasonable expectation of success goes to motivation or rationale as to why one of ordinary skill in the art would have found the claimed invention obvious. If the invention as claimed is not obvious, then an innovation can be patented." Strauss thought that the tree-related patents he cited not only failed to contain novel methods but also did not demonstrate effectiveness on a reasonable sample of the taxonomic groups for which claims were granted.

A closely related complaint was that scientific peer review plays no role in patent decisions. "If you ask scientists if they felt that a particular extension to a tree is really innovative or if it's just doing the same thing in a woody plant," Strauss said, "for some of these cases I've talked about they would say it's obviously the same thing. But the U.S. Patent and Trademark Office doesn't do that. They don't have a peer review process."

"You're absolutely right," Lee said. "We do not do peer review." Instead, the U.S. Patent and Trademark Office relies on examiners who are scientists, most of whom in the biotechnology area have Ph.D.s. They have 15 to 20 hours to make a decision on a particular patent application, and they must take into account not just scientific issues but legal ones as well. "Everything we do has a legal basis. We are scientists, but we must deal with the statutes. We deal with what the courts have determined to be patentable," Lee said. The awarding of a patent by the Patent Office is often just the first—and least expensive—round in deciding exactly what is covered by a patent. "The way patents get defined is that the U.S. Patent and Trademark Office issues something and then someone contests it in court," said John Reilly, an economist with the U.S. Department of Agriculture's Economic Research Service. "If no one contests it, it stands." Fighting it out in court does have its advantages. It allows the various interested parties to present their cases head to head, and a court has the time to make a more deliberate and well-thought-out decision than the U.S. Patent and Trademark Office can. But there are disadvantages too. A court battle can drag out for years, leaving everyone in limbo. And it can be so expensive that smaller companies may be effectively frozen out of the debate over how a patent should be defined.

oped a good technology "is going to get involved in some kind of a feud," and the legal dispute is likely to cost more than it did to develop the technology. Sometimes, he said, it is cheaper and easier to just buy the opposing company.

Summing up the effects of intellectual property on the agricultural biotechnology business Bedbrook said, "Right now it is a significant depressant in start-up and innovation in our industry because people cannot see their way clear to developing business." Steven Strauss of Oregon State, who leads a forestry consortium, echoed Bedbrook's assessment: "The scenario out there is an absolute mess. The complexity and the costs stifle small companies and small crops from getting into the business. We really have a monopoly situation developing, and I don't see it getting any better."

The issue of intellectual property rights in agricultural biotechnology looks quite different from the perspective of industry than it does from the viewpoint of academia. The reason is not hard to find, noted Ronald Sederoff of North Carolina State University, and it comes down to a difference in what Sederoff termed "means and ends." For academic institutions "money is the means to an end, and the end is that of acquiring scientific knowledge." But the ends and means are reversed for industry. "Companies want to make money. Scientific information is a means to make money, and money is the end." Because scientific information is simply a means to an end for them, companies have little of the devotion to open communication that permeates the university. Instead, they make their decisions about what and how much information to share with others based on the financial bottom line—a calculation that often leads to keeping as much information secret as possible. "Most major companies that I work with don't patent and don't disclose," said Robert Swank of the U.S. Environmental Protection Agency's National Exposure Research Laboratory. Ironically, most companies also recognize that their competitiveness in the future will depend on just the sort of creativity that flourishes in universities that enjoy open communication.

TECHNOLOGY TRANSFER

Referring to cooperation between academia and industry, Ronald Sederoff of North Carolina State University said, "It could be a marriage made in heaven—they give us money and we give them information. But it's a little more complicated than that."

Indeed, it has proved to be a lot more complicated than that. There was little that everyone at the forum agreed on, but one sentiment seemed to be almost universally shared: the current system of technology transfer from universities or government laboratories to industry needs to be greatly improved. The comments by Laura Meagher, associate dean of research at Cook College of Rutgers University, were representative of, if somewhat more diplomatic than, those of others: "I don't think we have managed to develop a good technology transfer system across the board in this country. There is a fair amount of lip service.

There are some outstanding pockets of effectiveness, but there are still some challenges ahead in implementation."

In theory, technology transfer should be straightforward. Universities and government agencies have a vast store of knowledge, inventions, and expertise that would be quite valuable to companies interested in developing commercial products, and since this intellectual property was generated with taxpayers' money, the idea is to put it to work in some way that will benefit the general public. The purpose of patenting and licensing inventions made with federal funding is to facilitate technology transfer to promote utilization of inventions coming from federally supported research and development. So universities and government laboratories sell their knowledge, inventions, and expertise to private companies, earning money for the institution and aiding the economy in a general way by helping industry create more and better products.

Unfortunately, it has not been as easy as it sounds. In the field of agricultural biotechnology, at least, technology transfer has been dogged by a number of problems, most of which stem from differences in how academia and industry operate, a lack of communication among the parties involved, and a failure to understand situations from the other side's perspective.

"One of the difficulties in licensing transactions is that it is very hard to value certain things," said John Bedbrook of DNA Plant Technology Corporation. "When we enter into a negotiation, we do not know how important this technology ultimately is going to be," and so no one knows for sure what price tag should be put on a particular bit of intellectual property. "When you are talking to another company, they have the same problem, so one can trade," Bedbrook noted. "Trading avoids the difficult question of quantifying the real value of a particular technology."

But, he said, "universities are not interested in trading technology. They are interested in receiving value for their technology in terms of compensation, or cash, or stock, or money, or some tangible asset. This forces the negotiation to deal with real value." And the need for putting a value on an invention makes negotiations concerning it much more difficult.

Furthermore, several industry representatives said, the process is made even harder by the tendency of university technology transfer offices to overvalue their intellectual property. As Robert Fincher explained, the value of a gene that could be inserted into a variety of corn or other seed crop tends to look much different to people at a university whose entire focus has been on that single gene than it does to their counterparts at a seed company:

> From the perspective of a seed company, we realize that the hybrid we put on the market is the product of two inbreds that have undergone maybe 10 years of breeding and testing, and they are dependent on previous breeding and testing and on however many genes there are in a corn plant—50,000 to 100,000, somewhere in that range. We feel that this is the package we are delivering to the farmer. Adding another gene to that package may or may not have the kind

> of incremental value that allows us to return some large percentage royalty to the university. Also, if we license, say, the structural gene from the university, we may still have to license the promoter from someone else, and the transformation system from someone else, and on and on. So that dilutes how much you can pay for the gene.

Besides overvaluing the technologies they have to offer, universities undervalue and misunderstand the risk that companies take in developing the intellectual property they license from universities, some forum participants said. "There can be a failure to make the distinction between an invention and a final product," said George Jen, an associate with Pennie and Edmonds LLP. "Not every invention leads to a good product, so there is considerable financial risk at the business end of investing and trying to develop an invention into a valuable product."

The tendency to undervalue risk can make technology transfer offices difficult to negotiate with, said Ilya Raskin, a plant physiologist at Rutgers University. Raskin described his own experiences with Phytotech, Inc., a company he founded to develop phytoremediation techniques—ways to use plants to clean up polluted soil and water. Phytotech, he said, "put $1.3 million in Rutgers for the phytoremediation research, taking all of this risk. The university position at that time was, 'While it may turn out to be good, and please give us the money, you also will be paying all the patent fees and all of the legal fees, everything. All of the risk is paid for by the company. The university just gets a straight research funding. Then if it works out well, we are going to negotiate on very good terms and give you all of the licenses you want. This will be easy.' "

"But it was not easy," Raskin said. Once Rutgers saw that phytoremediation was on its way to being a valuable technology, the university wanted "a certain amount of equity, significant royalties with the licensing, and so on." Eventually, it was resolved, Raskin said, but the events illustrate how little significance universities sometimes place on the risk that companies take.

These two biases—undervaluing the risk taken by companies and overestimating the value of their own inventions—were the most common complaints about technology transfer offices, but there were several others. Raskin, for instance, cited a "lack of motivation" at technology transfer offices, which leaves them unlikely to make the extra effort necessary to work out complicated arrangements with some companies. Other members of the forum found fault with technology transfer offices for having variable policies, so that companies never know what to expect when approaching a new university, and for not knowing exactly what their goals are in technology transfer. Businesses go into negotiations knowing what they want and how much they are willing to pay; universities often do not.

As was the case with concerns among forum participants about access to enabling technologies, large and small companies represented at the forum had very different opinions about technology transfer. While representatives from small companies generally complained about the efforts of university technology transfer programs, those from larger companies were much more satisfied with them. As Robert Fincher from Pioneer Hi-Bred International commented, "From what I have seen in working with the public sector and in working with the universities, we are able to exchange intellectual property and obtain intellectual property from each other fairly well." Wendy Choi, patent attorney at Union Camp Corporation, a forest products company, said, "Our experience has been very positive. We're getting information and work done that we haven't been able to get done before and it's been done by experts in the field. We're very pleased with it."

Choi's few complaints about working with universities centered on a failure to appreciate the importance of protecting intellectual property. "This lack of appreciation comes in a number of ways," she said. "If you don't understand what a patent is, it's hard to know when you're doing certain things that will preclude you from getting [one]." University researchers are often guilty of this, she said. Furthermore, scientists often fail to understand the need for delaying publication of certain research. "This is the single issue that is most often a stumbling block." The company understands researchers' need to communicate their findings, she said. "Believe me, I've written patent applications in two days and filed them because we knew somebody had to talk about it in two days." But the company does insist on being given the chance to file for a patent before researchers release their findings. Finally, Choi said, nondisclosure of the company's trade secrets often becomes an issue. "If we freely exchange the information with you so that it can help you do the research, we expect that that obligation of confidentiality is maintained," she said. But some researchers balk at signing an agreement to keep the company's confidential information out of their publications.

Such concerns, while important to small companies, are far from the most important, and Raskin suggested that technology transfer offices would be more effective if they realized this. "I think the offices of technology transfer need to understand the key differences in dealing with a large company and a small company," he said. "Small companies are equity rich and cash poor. The large companies have different problems. I work with Pioneer Hi-Bred and I work with Phytotech, and I have never seen any differentiation between those two done at the level of technology transfer offices."

The companies hurt most by the failings of the technology transfer system are the smaller ones. They have fewer resources to spend on negotiating with technology transfer offices and are less likely to employ someone with experience working these offices. They are less able to spend the necessary time—sometimes years—working out the details of an agreement, and they are not as willing to go to court to challenge a recalcitrant office. Indeed, Raskin said, many small businesses are so put off by the frustrations of working with technology transfer offices that they simply walk away. "Big companies may be used to the posturing and to the time which is required to get a decision," he said, but the smaller players are not, and sometimes they think they cannot afford to play the game.

THE FUTURE

Despite the difficulties that forum participants saw surrounding the issue of intellectual property rights in agricultural biotechnology, they also saw room for improvement and optimism. "A lot of what we're talking about is just growing pains," said Ronald Sederoff of North Carolina State University. "Much of what we're struggling through has been struggled through before in other disciplines, as territory became defined and as principles for patenting became tested, by either legislation or in the courts. We will be going through that same process. We will be traveling a similar path in our own special way."

Some things will get better with time, such as uncertainties concerning the patent rights to some of the enabling technologies in the field. Other things can be dealt with if people only realize what sorts of traps await and plan ahead to avoid them.

For example, Steven Strauss of Oregon State University noted that scientists in his forestry research consortium try to plan their work to minimize licensing complications down the road. "Even though we're doing research only right now, we try to foresee what some of the intellectual property complications and some of the licensing costs are going to be and to do research along lines that will lead to methods that will be [the] least onerous and most deployable when industries go to commercialize this."

But such thinking ahead demands that people become educated about the difficulties in this area and acquainted with ways to deal with them. To this end, several forum participants agreed with Ellen Friedman, a biochemist with the Biological Sciences Curriculum Study. She suggested that it would be useful to assemble some success stories—university-industry-government collaborations that work well, for instance—and perhaps collect some failure stories as well. Then others could learn what to emulate and what to avoid.

Robert Goodman, a plant pathologist, at the University of Wisconsin, noted that universities provide very uneven support for intellectual property protection. Even adequate intellectual property protection may not be available to some university researchers. Whereas industrial scientists receive high-quality technical support for intellectual property protection, university researchers often are "in the dark." This issue raises many concerns, including compliance with the Bayh-Dole Act.

Another issue that universities must sort out, said Ilya Raskin of Rutgers and Phytotech, Inc., is just what they want to get in exchange for their intellectual property. "There are really three basic ways in which a university can extract some value from its research. One of them, of course, is to take an equity position, if it is a start-up company, which is sometimes very appealing for the university. Another one is to ask for royalty returns and licensing fees. The third one is to try to get research contracts from the company as a partial compensation." Raskin argued that the third option is the natural one for universities to pursue. "Let us each do what we do best. Universities do research and education best. Companies do investing best."

On the issue of how university researchers deal with intellectual property rights, Sederoff suggested that, eventually, academics will have to learn to do things much as they are now done in industry. "In the end, I believe the industrial perspective will prevail. What industry is asking us to do in academia in fact is required in order to carry out the commercial aspects of the things we work on. And while I may have complained bitterly about it, I think that the perspective industrial scientists are promoting is actually correct. It's a completely different culture than some of us were raised in, but academic scientists are just going to have to become more comfortable with it and find ways to deal with it."

More generally, everyone agreed that universities are changing—and, indeed, must change—in response to various outside pressures. The federal government, for example, is pushing universities to do research with a greater "social return," that is, work that helps make life better in some way for the taxpayers who are paying for the research. But if universities are changing, the question becomes: Into what? Already there are signs that academia is moving toward an industry model in certain ways. Universities are, for instance, beginning to treat some of their intellectual property as bargaining chips, just as industry does.

But no one wants the distinction between research at universities and research in industry to disappear. There still is—and always will be—research that can only be done at universities, free of the pressure for short-term returns and against directions that offer no obvious potential for commercialization. Nonetheless, universities will almost certainly be asked to tie some of their research more closely to commercial purposes than has been the case in the past. With this in mind, Laura Meagher, associate dean of research at Rutgers, suggested that "we need some new sort of relationship between companies and universities, something that encourages long-term basic research in universities and yet encourages research that is conducted in general directions that will be of use to various industry sectors in the future."

This will be much easier, Meagher said, if the various players in agricultural biotechnology could learn to see the world from one another's perspective. "If we all saw ourselves as partners, university-industry-government, in this overall process of technology transfer, perhaps we could listen more to each other as to what our needs are and work out these win-win-win situations."

Appendixes

APPENDIX

A

Forum Program and Discussion Questions

Molecular biology has both transformed the science of biology and spawned the new industry of biotechnology. Biotechnology is a knowledge-based industry in which research breakthroughs are so broadly distributed that no single organization can keep pace with the information flow. Timely access to knowledge and resources that are otherwise unavailable are being provided increasingly through collaborative research projects involving industry, academe, and government.

The protection of intellectual property has arguably been the single most difficult issue to resolve in the realignment of partnerships between universities and industries in what is perhaps the most significant reorganization of basic research in the United States since Vannevar Bush. While stronger protection of intellectual property has increased incentives for private-sector investment in new agricultural technology, there is some concern that protective measures will impede the long-term progress of plant science. The forum will attempt to clarify the issues and explore possible solutions by examining controversies about intellectual property and its impact on plant biotechnology and fundamental research.

It will provide a neutral setting to promote the open exchange of views and will focus on three industrial applications of plant biotechnology: phytoremediation, biobased energy, and field crop breeding. To promote an intersector dialogue, panels of speakers from universities, industry, and federal agencies will examine issues and possible responses for each application. Discussion will be organized so as to compare and contrast issues among collaborators and across the plant sciences.

GOALS

The goals of the forum will be to bring together basic researchers in plant biology, policymakers, and those involved with the application of research in plant molecular biology. Participants will include research scientists; research and technology transfer managers, funders, and policymakers; product developers; and other experts in plant-based industries (especially crop genetics, biobased energy, and phytoremediation).

FORMAT

Speakers will discuss which intellectual property rights issues are the most important to address, which might be amenable to policy intervention, and whether further study of the issues and responses is warranted. The forum will provide a neutral setting to promote the open exchange of views. A summary report of the forum will be prepared for publication by the National Academy Press. The publication will not include recommendations. To encourage open discussion, no statements by speakers will be printed in the report without permission.

QUESTIONS

The following are among the questions relating to plant biotechnology that participants will consider:

1. *Is technology transfer managed in a way that provides for scientific progress, incentives for commercial development, and public benefit?* That is, are new techniques, information, research tools, and other forms of intellectual property effectively disseminated? If not, what improvements might be made?

2. *What benefits and problems result from negotiations or alliances between university, government, and commercial laboratories?* How do universities, government, and industry differ in their missions, motives, and expectations for collaborations? What licensing strategies and procedures for technology transfer will be most beneficial for the different sectors?

3. *Do current means for protecting intellectual property rights adequately encourage both scientific progress and commercial development?* Is the level of protection sufficiently strong to encourage commercial investment in the development of innovative products and techniques and yet sufficiently generous so that free exchange of scientific information is not impeded? Do current patenting practices need to be modified?

4. *What effect could current strategies for protection of intellectual property have on the progress of basic research and the future of plant biotechnology?* Is support for research into fundamental mechanisms adequately balanced with support for research for which commercial applications are closer at hand?

APPENDIX

B

Forum Agenda

Opening Remarks

Michael Clegg, *University of California, Riverside, and Chair, Board on Biology*
Bruce Alberts, *President, National Academy of Sciences*

Economic Overview

Suzanne Scotchmer, *University of California, Berkeley*

Panel One:
Crop Genetics

Robert Fincher, *Pioneer Hi-Bred International*
John Bedbrook, *DNA Plant Technology Corporation*
Patricia Swan, *Iowa State University*
Robert Goodman, *University of Wisconsin*
Alan Bennett, *University of California, Davis*
June Blalock, *Office of Technology Transfer, U.S. Department of Agriculture*

Panel Two:
Phytoremediation

Ilya Raskin, *Rutgers University/Phytotech, Inc.*
Laura Meagher, *Rutgers University/Phytotech, Inc.*
Robert Swank, *National Exposure Research Laboratory, U.S. Environmental Protection Agency*
Ellen Friedman, *Biological Sciences Curriculum Study*

Panel Three:
Biobased Energy

Gerald Tuskan, *Oak Ridge National Laboratory, U.S. Department of Energy*
Steven Strauss, *Oregon State University*
Wendy Choi, *Union Camp Corporation*
Ronald Sederoff, *North Carolina State University, and Member, Board on Biology*

Conclusions/Wrap-up

Michael Clegg, *University of California, Riverside*

APPENDIX

C

Speaker Biographies

Michael T. Clegg *(Forum Chair)* is acting dean of the College of Natural and Agricultural Sciences, University of California, Riverside. Dr. Clegg is the leading student of the evolution of complex genetic systems. He is recognized internationally for his contributions to the genetic and ecological bases for adaptive evolutionary changes within populations and at higher taxonomic levels. His current research interests include population genetics of plants, plant molecular evolution, statistical estimation of genetic parameters, plant phylogeny, plant genetic transmission and molecular genetics, and genetic conservation in agriculture. Dr. Clegg is a member of the National Academy of Sciences (NAS) and chair of the Board on Biology. He has served on several National Research Council studies and NAS commissions. Dr. Clegg received his Ph.D. in genetics from the University of California, Davis.

John R. Bedbrook is executive vice-president and director of science, DNA Plant Technology Corporation (Empressas La Moderna), Oakland, California. In 1980 he cofounded Advanced Genetic Sciences, Inc., which later merged with DNA Plant Technology Corporation. Dr. Bedbrook has published over 100 articles on molecular genetics. He was the first person to isolate a plant gene. He has served on the editorial boards of several international scientific journals and initiated the first teaching course in plant molecular biology at Cold Spring Harbor Laboratory. He has studied and worked in the field of plant molecular genetics at Harvard Medical School; Harvard University; the Plant Breeding Institute in Cambridge, England; and CSIRO in Australia. Dr. Bedbrook received his Ph.D. in molecular biology and virology in New Zealand.

Alan B. Bennett is professor and associate dean at the College of Agriculture and Environmental Sciences, Department of Vegetable Crops, University of California, Davis. Dr. Bennett's major research interests at the University of California include molecular biology of tomato fruit development, molecular basis of membrane transport, and protein maturation and targeting to the cell wall and vacuole. Dr. Bennett currently serves on the editorial board of *Plant Physiology* and has served as a panel member for the U.S. Department of Agriculture's competitive research grants and National Science Foundation programs. He also serves as the University of California representative on the National Agricultural Biotechnology Council. Dr. Bennett holds one patent (U.S. Patent #5,168,064) for Endo-1,4-b-glucanase gene and its use in plants and has applied for another (U.S. Patent Application #770,970) for tomato acid invertase gene. He received his Ph.D. in plant physiology from Cornell University.

June Blalock is a licensing specialist with the Office of Technology Transfer, Agricultural Research Service, U.S. Department of Agriculture, Beltsville, Maryland. Ms. Blalock joined the Office of Technology Transfer in 1993 as coordinator of the Technology Licensing Program. Previously, she was associate director of the Triangle Universities Licensing Consortium, where she had primary responsibility for licensing university-owned intellectual property in the biotechnology and biomedical fields from Duke University, North Carolina State University, and the University of North Carolina at Chapel Hill. She has held sales and marketing positions at International Biotechnologies, Inc., and has taught microbiology at the University of Maryland and Goucher College. Ms. Blalock is a member of the Licensing Executives Society, the Association of University Technology Managers, the Association of Federal Technology Transfer Executives, and the American Society for Microbiology.

Wendy A. Choi is a patent attorney with Union Camp Corporation, in Princeton, New Jersey. Before law school, Ms. Choi was a research scientist for the Rohm and Haas Company in Philadelphia. She graduated cum laude from Temple University School of Law and summa cum laude from Chestnut Hill College with a B.S. in chemistry. Ms. Choi is admitted to practice law in Pennsylvania and before the U.S. Patent and Trademark Office.

Robert R. Fincher is director of university-government research collaborations and germplasm licensing at Pioneer Hi-Bred International, Johnstown, Iowa. Previously, Dr. Fincher served as director of research for a group that works to improve agronomic traits with new technologies. The group's areas of expertise include breeding, genetics, statistics, molecular biology, and plant physiology. Dr. Fincher began his work at Pioneer Hi-Bred International in 1982 as a corn breeder and in 1985 began to work with biotechnology projects, including field evaluation of cell-culture-derived plants. He received his Ph.D. in agronomy (plant breeding) from the University of Missouri, Columbia.

B. Ellen Friedman directs a curriculum development project at the Biological Sciences Curriculum Study (BSCS), Colorado Springs, Colorado. She also designs and writes materials for a new college-level curriculum in biology being developed at BSCS. In academic settings, Dr. Friedman taught at the college and postgraduate levels and conducted research in molecular genetics, biochemistry, and genetic engineering during a 20-year period. Most recently, she was an assistant professor at New Mexico State University and was with the Plant Genetic Engineering Laboratory. She also served as a graduate faculty instructor for the molecular life sciences doctoral program at New Mexico State. Dr. Friedman received her Ph.D. in biochemistry from Rice University.

Robert M. Goodman is a professor in the Department of Plant Pathology, University of Wisconsin, Madison. He is also a member of the interdepartmental program in plant genetics and plant breeding, the Institute for Environmental Studies, the graduate program in cellular and molecular biology, and the Biotechnology Training Program. At the University of Wisconsin, Dr. Goodman's laboratory works on the molecular regulation of plant defense genes and the role of plant genotype in associations with noninvasive beneficial microorganisms. He is well known for his groundbreaking research as a professor at the University of Illinois, where he was the first to describe the molecular biology of a group of plant viruses, now called geminiviruses. Dr. Goodman has served on the National Research Council's Board on Agriculture and numerous NRC study committees. Dr. Goodman received his Ph.D. from Cornell University.

Laura R. Meagher is associate dean of research at Cook College and associate director of the New Jersey Agricultural Experiment Station, Rutgers University. She is also cofounder and director of Phytotech, Inc., of Monmouth, New Jersey. At Rutgers she has responsibility for leadership in catalysis and implementation of novel multidisciplinary multisector initiatives, such as the Biodiversity Center, and is involved in research and outreach in systematics, ecology, natural product chemistry, conservation and restoration, and an Ecosystem Policy Research Center, a soft-walled center that brings together teams of diverse social and natural scientists to address environmental, agricultural, marine, and science and technology issues. Previously, Dr. Meagher served as industry/government liaison for the Agricultural Biotechnology Center. In the early 1980s she was a cofounder and vice-president of the North Carolina Biotechnology Center. Dr. Meagher received her Ph.D. in zoology from Duke University.

Ilya Raskin is a professor at the Center for Agricultural Molecular Biology, Rutgers University, and founder, director, and chairman of the Science Advisory Board, Phytotech, Inc., of Monmouth, New Jersey. Previously, Dr. Raskin held positions at Dupont and Shell Agricultural Chemical Company. He has served on numerous review panels for the U.S. Department of Agriculture, National Institutes of Health, National Science Foundation, U.S. Department of Energy, BARD, Human Frontier Science Program, and AFRC (U.K.). Dr. Raskin has eight U.S. and foreign patent applications pending. He has two patents for (1) removal of metals from aqueous streams using plant roots (rhizofiltration) and (2) use of crop and crop-related species of the Brassiceae tribe of Brassicaceae family for in situ remediation of metal-contaminated soils (phytoextraction). In 1996 Rutgers University awarded him a Board of Trustees Award for Excellence in Research. Dr. Raskin received his Ph.D. in plant physiology from Michigan State University.

Suzanne A. Scotchmer is professor of economics and public policy, University of California, Berkeley. Dr. Scotchmer has eclectic academic interests that range from legal issues, such as intellectual property protection and rules of evidence in criminal trials, to evolutionary game theory. She has written on the process of jurisdiction formation, tax enforcement, and antitrust issues. She currently serves on the editorial boards of the *American Economics Review*, *Journal of Economic Perspectives*, *Journal of Public Economics*, and *Regional Science and Urban Economics*. She has been a visiting professor of economics at the new School of Economics in Moscow and at the Université de Paris I (Sorbonne), as well as Distinguished Olin Visiting Professor of Law and Economics at the University of Toronto. Previously, Dr. Scotchmer was associate professor of economics at Harvard University, Hoover national fellow at Stanford University, and Olin fellow at Yale Law School. Dr. Scotchmer received her Ph.D. in economics from the University of California, Berkeley.

Ronald Sederoff is Edwin F. Conger Professor of Forestry, Department of Forestry, North Carolina State University, Raleigh. Dr. Sederoff's research group leads the field in molecular genetics of forest trees, providing the base for genetic engineering of forest trees. His research group was the first to transfer a gene into a conifer. Dr. Sederoff and colleagues developed methods for genomic mapping of individual trees and applied those methods to complex trait analysis, particularly growth and disease resistance. Dr. Sederoff also is director of the North Carolina State University Forest Biotechnology Industrial Research Consortium. Dr. Sederoff is a member of the National Academy of Sciences and serves on the Board on Biology. He received his Ph.D. in zoology from the University of California in Los Angeles.

Steven H. Strauss is a professor of forest science, Oregon State University, Corvallis. Dr. Strauss's research topics include genetic engineering, genome mapping, and population genetics of forest trees. He also directs the Tree Genetic Engineering Research Cooperative at Oregon State University, a consortium composed of 11 paper companies, the U.S. Department of Energy, and the Electric Power Research Institute. He has held visiting scientist positions in France and Australia and currently serves as chairman of the International Union of Forestry Research Organizations Working Party on Molecular Genetics of Forest Trees. Dr. Strauss has obtained grants totaling over $3 million from the National Science Foundation, U.S. Department of Energy, U.S. Department of Agriculture, National Institutes of Health, U.S. Environmental Protection Agency, forest industries, and other sources. He has authored over 40 journal articles. He received his Ph.D. in forestry genetics and resource management from the University of California, Berkeley.

Patricia B. Swan is vice-provost for research and advanced studies and dean of the graduate school at Iowa State University in Ames. Previously, she was a professor in the Department of Food Science and Nutrition and associate dean of the graduate school at the University of Minnesota, where she had been a member of the faculty since 1964. Professor Swan's current work emphasizes the history of research on nutritional biochemistry in the United States. She has served on numerous professional societies and committees as well as the U.S. Department of Agriculture's Agricultural Advisory Committee on the National Research Initiative and as a member of the National Research Council's Board on Agriculture. She presently serves as a member of the Board of Directors for the Alternative Agricultural Research and Commercialization Center, as a member of the National Agricultural Biotechnology Council, and as president of the newly formed Iowa Research Council. Dr. Swan received her Ph.D. in biochemistry and nutrition from the University of Wisconsin.

Robert R. Swank, Jr., is director of research at the Ecosystems Research Division, National Exposure Research Laboratory, U.S. Environmental Protection Agency (EPA), in Athens, Georgia. Previously, Dr. Swank was director of research at the Athens Environmental Research Laboratory of the EPA. Dr. Swank is presently a member of the Science Advisory Committee of the South/Southwest Hazardous Substance Research Center. He has published several articles on industrial pollution control technology and exposure assessment methods. Dr. Swank received his Ph.D. in chemical engineering from the Georgia Institute of Technology.

Gerald A. Tuskan is a research staff scientist with Oak Ridge National Laboratory, Environmental Sciences Division, U.S. Department of Energy, in Oak Ridge, Tennessee. His research responsibilities are in the areas of molecular/genetic diversity studies of newly regenerated aspen seedlings in Yellowstone National Park; genetic transformation experiments involving plant hormone genes; and genetic characterization of stress resistance in woody plants, particularly temperature adjustment, drought, and UV-B stresses. Previously, Dr. Tuskan was an assistant professor at North Dakota State University. He has authored over 48 refereed publications and 32 papers. Dr. Tuskan received his Ph.D. in forest genetics from Texas A&M University.